EDUCATING ENGINEERS

Preparing 21st Century Leaders in the Context of New Modes of Learning

SUMMARY OF A FORUM

Prepared by Steve Olson
for the
NATIONAL ACADEMY OF ENGINEERING
OF THE NATIONAL ACADEMIES

I0807145

THE NATIONAL ACADEMIES PRESS
Washington, D.C.
www.nap.edu

THE NATIONAL ACADEMIES PRESS 500 Fifth Street, NW Washington, DC 20001

NOTICE: The subject of this report is a forum titled Educating Engineers: Preparing 21st Century Leaders in the Context of New Modes of Learning held during the 2012 Annual Meeting of the National Academy of Engineering.

Opinions, finding, and conclusions expressed in this publication are those of the workshop participants and not necessarily the views of the National Academy of Engineering

International Standard Book Number 13: 978-0-309-26770-0
International Standard Book Number 10: 0-309-26770-6

Copies of this report are available from the National Academies Press, 500 Fifth Street, NW, Keck 360, Washington, DC 20001; (800) 624-6242 or (202) 334-3313; http://www.nap.edu.

For more information about the National Academy of Engineering, visit the NAE home page at **www.nae.edu**.

Copyright 2013 by the National Academies. All rights reserved.

Printed in the United States of America

THE NATIONAL ACADEMIES

Advisers to the Nation on Science, Engineering, and Medicine

The **National Academy of Sciences** is a private, nonprofit, self-perpetuating society of distinguished scholars engaged in scientific and engineering research, dedicated to the furtherance of science and technology and to their use for the general welfare. Upon the authority of the charter granted to it by the Congress in 1863, the Academy has a mandate that requires it to advise the federal government on scientific and technical matters. Dr. Ralph J. Cicerone is president of the National Academy of Sciences.

The **National Academy of Engineering** was established in 1964, under the charter of the National Academy of Sciences, as a parallel organization of outstanding engineers. It is autonomous in its administration and in the selection of its members, sharing with the National Academy of Sciences the responsibility for advising the federal government. The National Academy of Engineering also sponsors engineering programs aimed at meeting national needs, encourages education and research, and recognizes the superior achievements of engineers. Dr. Charles M. Vest is president of the National Academy of Engineering.

The **Institute of Medicine** was established in 1970 by the National Academy of Sciences to secure the services of eminent members of appropriate professions in the examination of policy matters pertaining to the health of the public. The Institute acts under the responsibility given to the National Academy of Sciences by its congressional charter to be an adviser to the federal government and, upon its own initiative, to identify issues of medical care, research, and education. Dr. Harvey V. Fineberg is president of the Institute of Medicine.

The **National Research Council** was organized by the National Academy of Sciences in 1916 to associate the broad community of science and technology with the Academy's purposes of furthering knowledge and advising the federal government. Functioning in accordance with general policies determined by the Academy, the Council has become the principal operating agency of both the National Academy of Sciences and the National Academy of Engineering in providing services to the government, the public, and the scientific and engineering communities. The Council is administered jointly by both Academies and the Institute of Medicine. Dr. Ralph J. Cicerone and Dr. Charles M. Vest are chair and vice chair, respectively, of the National Research Council.

www.national-academies.org

Preface

Every year the National Academy of Engineering holds a three-hour forum at its annual meeting on a subject of pressing importance, and in 2012 the forum was on educating engineers for the 21st century. I am deeply interested in education and have written and spoken often on the subject during my six years as NAE President, but my enthusiasm for the topic was not universally shared before the forum. "We have been talking about education for decades, but very little has changed," was one comment made at the forum. "What more can we say about education than has already been said?"

By the end of the forum the tone had completely changed—I heard nothing but excitement. The vision laid out by the six speakers was breathtaking. They described a future that marks a profound break with the past. The ongoing revolution in computing and communications is part of the story, but the transformation is institutional and societal as well. New kinds of colleges and universities have been designed to take advantage of the unique opportunities created by the confluence of new technologies, modern markets, and social needs. Minorities now account for 40 percent of college-aged Americans, which means that the engineering community will soon be much more diverse than it has been in the past. A new generation of leaders in education and industry has emerged with the courage to do things differently, measure the results, and find out what works.

Certainly one reason for the success of the 2012 forum was the preeminence of the speakers. Tuula Teeri, founding president of Aalto University in Helsinki, and Rick Miller, founding president of Olin College outside Boston, described the tremendous advances that can be achieved through a reenvisioning of higher education. Salman Khan, founder of the Khan Academy, and Anant Agarwal, who organized the first massive online open course at MIT, pointed to the revolution in learning being driven by interactive online learning. Rick Stephens, vice president of human relations at the Boeing Company, and Linda Katehi, chancellor of the University of California, Davis, detailed steps that need to be taken for America to remain at the forefront of technology and business. All six speakers were provocative, visionary, and inspiring.

The forum was moderated, as it has been for the previous three years, by Ali Velshi, chief business correspondent for CNN. During his introductory comments, Ali mentioned writing a magazine article entitled "If I Could Do It All Over Again, I'd Be an Engineer," which is just one measure of how fortunate the NAE has been to have Ali as a partner and proponent. He has done a wonderful job of describing the challenge and excitement of engineering to his viewers, and we are grateful for his association with the NAE.

Since the NAE began holding these forums three years ago, I have noticed a new urgency surrounding the topics discussed. A change is coming to engineering that will be impossible to avoid. Engineering is becoming faster, more complex, and more global. Young engineers sense that they have an obligation to society to be entrepreneurial and to help develop the economies of regions, nations, and the world. Engineers are being drawn into new kinds of collaborations with designers, social scientists, business leaders, and educators to address difficult problems and exciting new opportunities.

No one knows exactly how engineering will change in the future, just as no one knows exactly how information technologies will change society or how engineers will be educated. But the 2012 forum provided one of the best glimpses we're likely to get of where engineering is headed and how our educational institutions will need to change to get us there.

Charles M. Vest, President
National Academy of Engineering

Contents

1

A Vision of the Future

The National Academy of Engineering's 2012 forum, "Educating Engineers: Preparing 21st Century Leaders in the Context of New Modes of Learning," opened with presentations by six speakers who looked at the future of engineering and engineering education from their perspectives as educators, administrators, entrepreneurs, and innovators. Each speaker focused on just one facet of a tremendously complex picture. Yet together they outlined a new vision for engineering education based on flexible, interactive, lifelong learning and the merger of activities long held to be distinct.

REVISIONING ENGINEERING EDUCATION: OLIN COLLEGE

When Richard Miller, the president of the Franklin W. Olin College of Engineering, came to the institution in 1999, he was the college's first employee. The College did not yet have buildings, faculty members, or students. What it had was a goal, said Miller: to rethink engineering education from the ground up.

The college began by assembling a group of founders who spent two years investigating what engineering is, how people learn, and how to restructure engineering education (see Box 1). They agreed that engineering is not a body of knowledge. "Engineering is a process. It's a way of thinking," said Miller. "The aircraft industry started in a bicycle shop. It didn't start with folks who had a PhD in physics." For the definition of an engineer, the founding group decided on the following: an engineer is a person who envisions what has never been and does whatever it takes to make it happen.

Box 1
A New Model for Engineering Education

In 1997 the F.W. Olin Foundation, established in 1938 by engineer, entrepreneur, and philanthropist Franklin W. Olin, announced its intention to create a new engineering college based on an innovative model of engineering education. The college began by conducting "Invention 2000," a two-year effort to rethink engineering education and college operations, then constructed a 300,000-square-foot campus with state-of-the-art academic, administrative, and residential facilities. The college received 2,000 applications for its 20 academic positions. Before the college opened, faculty members worked with 30 student partners who came to the college for a pre-freshman year to help create the curriculum and develop student life programs.

In August 2002, 75 students entered as Olin's inaugural freshman class. Initially, students paid no tuition. They now pay half of the tuition unless they have financial need, in which case they pay less. Faculty members are not tenured and are on renewable contracts, and the college has no academic departments. Students spend a substantial portion of their time doing projects in interdisciplinary courses that emphasize teamwork and communication skills. The curriculum combines a rigorous engineering education with entrepreneurship, the arts and humanities, and the social sciences, with an overall goal of creating well-rounded "Renaissance engineers."

For more information: *www.olin.edu*

Olin College does not want its students to have to know what scientific or engineering discipline they are learning. It wants students to be solving real problems from the day they arrive. Process needs to be at the center of engineering education, with science the scaffolding around that process to help students achieve high results.

"Engineering, in a way, is a performing art," said Miller. "Suppose you had a child who really wanted to be a violinist and went off to a conservatory of music. If that conservatory of music had an educational program that was patterned after the way we teach engineers today in most engineering schools, what would happen? In the first year, the student would take a course in the theory of sound. We would talk about vibrations. We would do physics problems. We would figure out how strings vibrate, what modal shapes are, and frequencies. In the second year we might take a course in the theory of composition—harmonics,

Richard K. Miller, president, Olin College

melodies, and so on. If you're really patient, in the fourth year, they might allow you to touch the violin and actually play some scales. I don't know a kid who is a prodigy in violin who would wait until the fourth year to do that."

The 21st century will be characterized by complexity, said Miller. "The problems are no longer contained in one continent. They transcend time zones. They transcend political boundaries. They transcend disciplines. They are no longer technology problems. They are societal problems."

> "We deliberately want to mix the DNA of engineering students and entrepreneurial business students."
> *Richard Miller, Olin College*

Yet engineering education today often works against this reality. Students who choose to be engineers spend four years with other engineering students rather than interacting with the students from other disciplines who will be essential to solving societal problems. Olin College, which is in Needham, Massachusetts, just outside Boston, is located next to Babson College, which has been ranked number one in entrepreneurship, and every student at Olin has to be involved in business to graduate. "We deliberately want to mix the DNA of engineering students and entrepreneurial business students."

The college emphasizes students' engagement in their own learning. Innovations in education, whether online learning or work-based experiences, are beneficial only to the extent that they result in increased student engagement, said Miller. The college also embraces the idea that people learn and succeed in many different ways.

"It takes more than SAT tests and mathematics and physics in order to be successful," said Miller. Other qualities can spur success, such as interpersonal intelligence and creativity. Many students have these qualities, and these students have the potential to become leading engineers. "We just tend to ignore them in traditional schools." Olin College has been working hard to bring people into the engineering profession who would not normally become engineers. "A lot of the kids who apply to our program don't apply to any other engineering school in the country. But the way we are thinking about engineering is attracting them." Miller observed that half of the incoming students are women.

Finally, a major goal of Olin College is to advance engineering education in the United States and throughout the world. Miller said that he talks frequently with other people who are in the process of inventing new universities. "If we come up with a good idea that happens to work, our mission is to give it away."

A NEW ROUTE TO COMPETITIVENESS: AALTO UNIVERSITY

One motivating force behind the creation of Aalto University was that "technology alone does not sell products anymore," said Tuula Teeri, the university's president. Instead, products need to integrate engineering, design, and market considerations to capture the interest of consumers. Apple has become dominant because it has understood what people want and has designed its products to meet those preferences, Teeri observed. Aalto University is building on this insight to educate the future creators of world-class products (see Box 2).

Finland is a country of just five million people, so it must look beyond its borders to ensure its prosperity. One way to increase its competitiveness is to emphasize entrepreneurship, which is at the center of Aalto University's mission. "What we have found out, in this short history of our new format, is that our students are powerful entrepreneurs—they have a huge capacity, much more than my generation.... In

Box 2
Integrating Institutions to Integrate Knowledge

The Aalto University School of Science and Technology in Finland was created in 2010 through the merger of three existing universities: the Helsinki University of Technology, the Helsinki School of Economics, and the University of Art and Design Helsinki. The university, which enrolls about 20,000 students, is organized into six schools—science, engineering, electrical engineering, business, chemical technology, and the arts, design, and architecture—but its underlying objective is the integration of knowledge. The merger of institutions provides abundant opportunities for multidisciplinary education. Similarly, research at the university is focused on themes that require a cross-disciplinary approach.

The university is named after Alvar Aalto (1898–1976), a Finnish architect, designer, and entrepreneur who exemplified the spirit of integrated knowledge on which Aalto University is based.

For more information: *www.aalto.fi/en*

engineering education, we should begin to give a lot more responsibility to our students, because, after all, they are the ones who are going to build our future."

Engineering education in Finland is traditionally a five-year program, where students begin studying engineering on their first day in the program and continue doing so until their last day. Aalto University has been changing this formula by thinking about the skills that future engineers are going to need. "It's very difficult to predict the future," said Teeri. "We don't really know what the jobs are that our engineers are going to take when they are finished [with their undergraduate education] and throughout their careers. Therefore, we are trying to go away from this kind of very tightly disciplinary education in engineering and make it broader, more multidisciplinary."

"Our students are powerful entrepreneurs—they have a huge capacity, much more than my generation.... In engineering education, we should begin to give a lot more responsibility to our students, because, after all, they are the ones who are going to build our future."

Tuula Teeri, Aalto University

Tuula Teeri, president, Aalto University, Finland.

In the future, engineers will learn less from books and lectures and more from other kinds of forums, including the Internet, said Teeri. The university will still play a critical role in organizing and overseeing this information, since online information is not always accurate, and in teaching students to be critical thinkers. The university also will promote student learning by engaging companies and other employers.

In particular, Aalto University emphasizes collaborations with industry. Students are supervised by academic teachers and industrial representatives simultaneously and work on ideas that come predominantly from industry. An industrial researcher works with a university team for six months, and many industrial liaisons find the experience immensely rewarding. "One of the engineers who had been working for 20 years in industry said this had been the most exciting experience of his life, because he is able to share with the students the experiences that he has had in industrial life for 20 years—and not just one student, but a whole cohort of students."

AN INVESTMENT IN THE FUTURE: THE KHAN ACADEMY

When an engineer friend suggested that Salman Khan put the tutorials he had been developing for his cousins on YouTube, Khan was skeptical—"I said, 'YouTube is for cats playing piano'"—but he decided to give it a try. "My cousins, after I had produced 20 or 30 of these… somewhat famously told me that they liked me better on YouTube than in person. I took that as positive feedback, and I kept going."

The YouTube videos proved to have two major advantages, Khan said. They were timeless in that many of the ideas Khan was describing were hundreds of years old. And they could scale up, because anyone in the world could watch them. After a few months, he realized that people who were not his cousins were watching. High school and college students were contacting him and saying, "This helped me get an A on my exam." People who had never understood a concept or who had missed a day in class were able to learn the concept and move on.

In 2008, Khan established the not-for-profit Khan Academy with the mission of "a free world-class education for anyone anywhere" (see Box 3). By 2009, traffic on the site was so great that he quit his job at a hedge fund firm to focus on what he hoped would be an investment with

Salman Khan, founder, Khan Academy.

Box 3
From a Classroom of One to a Classroom of Millions

In 2004, Salman Khan, a hedge fund analyst with degrees from MIT in mathematics, electrical engineering, and computer science, began tutoring his cousin in mathematics. Because she lived on the other side of the country, he talked to her over the phone while scrawling equations on a drawing program that she could watch on her computer. Sometimes he would record a lesson as a video and send it to her, and soon several other cousins were following his lessons as well.

From these modest beginnings grew an online educational initiative that now reaches millions of students every month. With funding from foundations, companies, and individual entrepreneurs, the Khan Academy provides free access to thousands of online videos, most of which Khan has recorded himself. Analytic software tracks students' use of the videos and their progress in mastering the content. Schools are experimenting with the use of Khan's videos to provide instruction in classrooms and at home while teachers focus on targeted interventions and special projects, and the videos are being translated into other major languages.

For more information: *www.khanacademy.org*

a very high rate of social return. The next year, several prominent companies, foundations, and capital investment firms began to invest in the project, and Khan started hiring "some of the smartest people I knew to build out this thing." The Khan Academy reached 7 million students in the month before the forum, and its growth remains explosive. "We keep trying to push the envelope on what we can be."

> "Engineering is just as creative as being an architect or a designer, because it's all about [developing] a portfolio of creative works."
> *Salman Khan, Khan Academy*

The Khan Academy is essentially a virtual school. Though built around interactive videos, it is designed to foster communities of users, and its rich simulations and data analytics optimize engagement. As a complement to a physical school, the Khan Academy provides resources that many students cannot get anywhere else.

The success of online instruction also makes it possible to rethink education, Khan said. Every student could work at his or her own pace, mastering each concept in turn before moving on. The role of the teacher would not be to lecture but to analyze data and do focused interventions based on the results. Teachers could focus on projects, including engineering projects, from an early age. "That has inserted us into the conversation of how we rethink what a classroom could be, what an education could be, what a credential could be."

In the future, online tools could accomplish much of what happens in traditional classrooms, Khan continued. Today, schools tend to focus on rote procedures and information transfer more than on creativity and teamwork. If technology could develop and assess the competency of students in core areas of knowledge, schools could be more project based and open ended. Such a reorganization would "make it clear that engineering is just as creative as being an architect or a designer, because it's all about [developing] a portfolio of creative works."

SCALING UP ONLINE EDUCATION: EDX

In the spring of 2012, a joint venture of Harvard and MIT called edX offered an open online course entitled "Circuits and Electronics." More than 150,000 students worldwide signed up to take the course, even though it required knowledge of advanced calculus and complex analysis, and more than 7,000 students passed it. "That's as many students as would take the class at MIT in 40 years," said Anant Agarwal, the president of edX and a professor of electrical engineering and computer science at MIT. "We taught that class with about the same level of staffing as we would teach a one-semester course at MIT, which about 100 to 200 students take.... It's pretty staggering."

Massive online open courses, or MOOCs, are "the biggest innovation in learning since the printing press," according to Agarwal (see Box 4). "Bringing technology to learning and applying it in a concerted manner would truly revolutionize the world."

EdX, which now involves a number of other universities in addition to Harvard and MIT, mixes elements of traditional classes with new capabilities made possible through technology. Students watch short videos that Agarwal called KSVs, for Khan-style videos (at the forum, Khan said that he had never heard of KSVs but was flattered by the term), and then engage in an interactive experience, thus requiring the students to do more than just watch a video. Together the vid-

Box 4
The Dawn of the MOOC

EdX is one of several new initiatives that have recently been established to offer massive online open courses, or MOOCs, to many thousands of students at a time. MOOCs are in their earliest stages, but they have the potential to create massive disruptions in higher education. Students who take such courses do not yet receive college credit, but plans are under way to conduct secure and credible assessments of what students taking these courses have learned. By monitoring students' use of online materials, researchers can learn more about how to improve online learning. Colleges and universities could begin relying on MOOCs for lectures while specializing in more personalized, hands-on experiences. Online courses could break down the current pricing structure of college and radically enhance the internationalization of education.

For more information: *www.edx.org*

eos and interactive experiences make up a learning sequence. Students who complete the course receive a certification from MITx, or from Harvardx, or from the other universities that are joining edX. EdX also has announced a partnership with the Pearson company that will allow students to take proctored exams throughout the world to demonstrate their mastery of the material.

One challenge for edX, said Agarwal, is to foster creativity in its students, not just engagement with the material. Toward that end, it has been creating online laboratories in which students can manipulate components on a screen to build devices and conduct experiments. Electrical engineers, for example, might be given electrical components and told to build a device, while civil engineers might be given members and beams. The laboratories even have music that students can incorporate into their designs.

> "Bringing technology to learning and applying it in a concerted manner would truly revolutionize the world."
>
> *Anant Agarwal, edX and MIT*

EdX is also seeking to encourage creativity by building community among online students. For example, it has structured its online laboratories as wikis where students from all over the world can contribute to

Anant Agarwal, president, edX, and professor, Massachusetts Institute of Technology.

a design. "People around the world are modifying each other's stuff," said Agarwal. "What we are trying to do is bring the creativity and the community into it in a big way."

PROMOTING K–12 ENGINEERING EDUCATION AT THE NATIONAL ACADEMY OF ENGINEERING

Linda Katehi, chancellor of the University of California, Davis, has always been acutely aware of how few women and minorities pursue engineering. When she was a first-year college student in Greece in 1972, she was one of only two women in a class of 190. When she went to the University of California, Los Angeles, as a graduate student in electrical engineering, she was again one of very few women. Then, when she became a faculty member at the University of Michigan, she was one of three female faculty members in a department of about 95 faculty. And the lack of minority students in engineering has been just as striking as the lack of women.

This underrepresentation reflects a more general neglect of engineering in the United States, she said. In China and in Europe, 21 per-

Linda P.B. Katehi, chancellor, University of California, Davis

cent and 14 percent of undergraduate degrees, respectively, are in engineering, compared to only 4 percent in the United States.

The best way to change this statistic is to begin teaching engineering very early in children's learning experiences—even before kindergarten, said Katehi, who chaired the NAE Committee on K–12 Engineering Education (see Box 5). Engineering is not just a collection of information. It is the ability to be observant, to identify problems, to be creative, to think about solutions, and to have the skills to realize solutions. The brain learns by putting things together. A favorite toy among children remains Legos, but children do not get to play with Legos in school. Instead of making things, schools provide students with information. Schools need to take advantage of how the brain works by allowing students to pursue their curiosity and connect that curiosity with science, said Katehi.

Box 5
K–12 Engineering Education as a Catalyst of Reform

Engineering has slowly been making its way into K–12 classrooms in the United States. A 2009 report from the NAE Committee on K–12 Engineering Education noted that thousands of teachers have attended professional development sessions on engineering-related coursework, and millions of K–12 students have experienced some formal engineering education.

K–12 engineering education can boost student achievement in science and mathematics, raise awareness of engineering and the work of engineers, and increase interest in engineering careers, the committee stated. Engineering education could even act as a catalyst for a more interconnected and effective system of K–12 education in science, technology, engineering, and mathematics (STEM). For example, scientific investigation and engineering design are closely related activities that can be mutually reinforcing, and mathematical analysis and modeling are essential to engineering design.

For more information: *www.nap.edu/catalog.php?record_id=12635*

The recently developed frameworks for science standards acknowledge this point by recognizing that learning science is best achieved through practice. "Engineering is practice," said Katehi. "It's designing things that can improve our quality of life, things that can solve some of our problems."

> "Engineering is practice. It's designing things that can improve our quality of life, things that can solve some of our problems."
>
> *Linda Katehi, University of California, Davis*

Engineering provides a way to rethink K–12 math and science education to show students how it becomes interesting and useful through engineering. Enhanced K–12 science education can attract the interest of more girls and minority students by appealing both to their curiosity and to their concern about the social impacts of engineering.

Engineering is becoming more global and socially conscious, Katehi concluded, and in an increasingly "flat" world, where technologies quickly diffuse around the globe, the United States needs to draw on its entire human resource base to retain a leadership position in both science and engineering.

WHAT INDUSTRY NEEDS: THE BOEING COMPANY

In 2011 the Boeing Company hired 18,000 employees, including thousands of engineers, and it plans to hire many thousands more engineers as its baby boomer employees retire. "In an environment where unemployment in engineering is about 2.6 percent, our worry is, Will there be enough talent to replace those who are currently providing the leading edge of technology that we need as a nation?" said Richard Stephens, senior vice president for human resources and administration for Boeing (see Box 6).

The problems that need to be solved by industry are ever more complex, whether they involve designing a water system, a hospital, a road, or an airplane. As such, many considerations—not just technology—need to be integrated to solve these problems, including social considerations such as responsibility for the environment. Boeing is interested in engineers who can help achieve this integration.

Stephens acknowledged that engineering schools already produce technically competent people, but argued that Boeing needs engineers who are more than just technically competent. First, it needs engineers who are creative. Sometimes engineering requires straightforward approaches, but many solutions are not predetermined. Technology will be important in a "nondeterministic world," said Stephens, to spur and support creativity.

Boeing also needs engineers who can work in teams. "The ability to communicate, interact with others, be on a team, share thoughts and ideas, have great discussions [is] critically important. And I would contend that one of the challenges we're beginning to see more and more

Box 6
The First Century of the Boeing Company

The Boeing Company was founded in 1916 in Seattle as a builder of wooden seaplanes. Today it has 175,000 employees and is one of the largest aerospace companies in the world. During its history, it has produced a large collection of successful commercial and defense aircraft, helicopters, missiles, rocket stages, satellites, spacecraft, parts of the International Space Station, and even hydrofoils.

For more information: *www.boeing.com*

Richard (Rick) Stephens, senior vice president for human resources and administration, the Boeing Company.

is that our youngsters grow up in the digital world and they are less and less interested and less and less adept at having real human contact and interaction. I think that's a key challenge we face."

Finally, Boeing needs engineers who will pursue new knowledge throughout their careers. This drive starts from an early age. Stephens told the story of a chief engineer at Boeing whose father hung a model of an F-15 from his ceiling when he was a child, and he started becoming an engineer as he wondered what it would be like to build airplanes like the F-15. Similarly, many engineers who are nearing the end of their careers were inspired as students by the challenge of Sputnik and reclaiming the American lead in aerospace.

Producing engineers with these traits will require providing students with role models and mentors to help them through the first few years.

Also, disciplines such as mathematics or physics should be taught in context, said Stephens, not as abstract entities, even if that means having engineers teach those subjects rather than people from those disciplines. Engineering schools should involve students in projects from day one, because engineering students want to solve problems, and projects give them a way to do so while learning about partnerships, relationships, and exchanging ideas, Stephens observed. Finally, engineering schools should ensure that their students have internships between at least their sophomore and junior years. Companies like Boeing need to support such internships, said Stephens, "because we're the ones that want the real, hands-on, practical experience." Boeing doubled its summer engineering internships to 1,200 in 2012, and plans to support more in future years.

> "In an environment where unemployment in engineering is about 2.6 percent, our worry is, Will there be enough talent to replace those who are currently providing the leading edge of technology that we need as a nation?"
>
> *Richard Stephens, Boeing Company*

Some contend that few US students want to become engineers because the subject is too hard. Stephens strongly disagreed with that idea. Some engineering schools are graduating more than 80 percent of their entering first-year students within six years. These schools could serve as models for how to make engineering attractive again.

II

Applying the Vision

During the second half of the forum, the speakers addressed questions posed by Velshi and members of the audience. Many of the questions concerned familiar topics such as interdisciplinary education, the roles of teachers, and the future supply of engineers, but the visions laid out in the speakers' presentations contributed to fresh and invigorating answers that transcend past approaches.

HOW LONG SHOULD IT TAKE TO EARN AN ENGINEERING DEGREE?

A forum participant pointed out that, in Chile, engineering is a six-year degree, and engineers have more prestige and are more highly paid than doctors and lawyers. But sentiment is growing in Chile for a four-year program in engineering, as in the United States, with a fraction of students going on to the master's and doctoral levels.

Khan thought that good engineers could graduate from college in less than four years, so long as they are exposed to good projects and develop the can-do attitude that top engineering students have. "I don't think it's a matter of seat time. It's a matter of competency and having a portfolio of works."

Stephens raised the question of how people in industry know that they are hiring good engineers. Portfolios and internship performances are indications of quality, but Boeing has to hire thousands of engineers, and those engineers need to be able to make airplanes that will fly.

Khan agreed with that observation. The problem, he said, is that degrees are not necessarily a good indicator of quality. To hire engineers, he relies on interviews that have both a technical and a social compo-

CNN chief business correspondent Ali Velshi and forum panelists.

nent. The great advantage of online learning, he said, is that it will shift educational outcomes toward measures of competency, not just seat time or grade point averages. Furthermore, competency-based skills could be either academic or social, based on the kind of online learning undertaken. Even students who specialized in other areas and then discovered a love of engineering could learn skills online "and then prove that they are just as good as the MIT grad or the Stanford grad," which in addition would help address projected shortages of engineers.

> "I don't think it's a matter of seat time. It's a matter of competency and having a portfolio of works."
> *Salman Khan, Khan Academy*

Continuing education will be critical in a world where knowledge is growing so quickly. Stephens said that the Boeing Company spends $100 million on sending 22,000 of its employees to school, and overall it spends three-quarters of a billion dollars on training and education. "We believe in it that much."

Agarwal, drawing on an idea from a recent conversation with an MIT alumnus, described a model of education that relies heavily on continuing education. Instead of having engineers complete a four-, five-, or six-year degree program, why not bring students into universities for a year or two to orient them to engineering and point them in the right direction? Then let them loose into the world to practice engineering while they continue to learn online. That way, even as they are learning online, they would be learning by doing.

SHOULD ENGINEERING EDUCATION BECOME MORE INTERDISCIPLINARY?

In response to a question about the structure of the engineering curriculum, Agarwal called attention to the strict disciplinary boundaries that persist in most universities. Departments tend to dwell on the needs of the department more than the needs of students. At MIT, for example, students from mechanical engineering who want to take a popular electrical engineering course cannot do so because the two courses traditionally have been taught at the same time. Today's students "are very interdisciplinary," he said. "They want to have much more of an international experience, much more of a multidisciplinary experience, but our departments are very stovepiped." Universities need to find ways to deliver broad interdisciplinary degrees to serve students who seek an interdisciplinary experience, he observed.

One way to promote interdisciplinary education, said Miller, is to put people back into textbooks along with scientific principles and engineering problems. Reynolds and Mach were people, not just numbers. People learn the most from stories, said Miller, and especially stories about people. At Olin College, a course called "The Stuff of History" combines the history of science and the principles of materials science. The course is team-taught by a historian and a materials scientist, and they cite the life of Paul Revere, who was a metallurgist and an entrepreneur as well as a patriot. "People will remember that. The students will have it in context."

> "They want to have much more of an international experience, much more of a multidisciplinary experience, but our departments are very stovepiped."
>
> *Anant Agarwal, edX and MIT*

Miller told a story about the Stanford economist Paul Romer's work on "charter cities"—cities designed and built essentially from scratch to take advantage of 21st century opportunities while serving the objectives and meeting the needs of their occupants. Olin College put together an interdisciplinary team of about 30 students, including students from

Wellesley College and Babson College, who spent a month developing a blueprint for such a city. They addressed problems involving transportation, sustainable energy, social justice, banking, international relations, and many other issues. The experience "changed the way they think about their careers and their ability to interact and to work on big problems outside of their disciplines."

Teeri reminded the group that a multidisciplinary education in engineering must still give students a deep understanding of something. When students are building bridges and aircraft, they need a science-based curriculum. Aalto University tries to provide its students with a broad-based literacy at the undergraduate level and then lets them choose an in-depth topic of study at the graduate level. Students often have a better sense than do faculty members about the kinds of jobs that will be available in the future, jobs that in many cases will be unlike today's jobs. Also, because the industries of the future cannot be predicted, engineers need a broad education while also being able to learn and practice skills on the job. "Learning to learn is one of the key things for future engineers."

Universities consider faculty to be the core of their institutions, said Katehi, whereas in fact students are the core. If curricula were redesigned around the needs of students rather than around the needs of faculty members, they would look quite different. Today, faculty members expect and want to deliver courses in their own areas of expertise. "If we put the student at the center and if we start looking at everything through that lens, then our decisions will be totally different."

IS THE UNITED STATES PRODUCING ENOUGH ENGINEERS?

Several panel members observed that the United States is not producing enough engineers to meet future needs, particularly in high-demand fields. "If we look at the demographics across industry, we have too many people who are going to retire," said Stephens. "We don't see a pipeline large enough to be able to fill that need."

> "If we look at the demographics across industry, we have too many people who are going to retire. We don't see a pipeline large enough to be able to fill that need."
>
> *Richard Stephens, Boeing Company*

Katehi emphasized the need for diversity if America is going to produce enough engineers. That means getting women and students of color interested in engineering from an early age, which has the additional benefit of spurring creativity by bringing additional perspectives to the field.

Agarwal called particular attention to computer science, where the shortage of students is "a disaster." He said, "We have a huge number of unfilled openings, and we graduate very, very few."

Khan pointed to an interesting development, which is that incomes have been rising rapidly for college graduates with computer science degrees. "A good trend is that those computer science incomes are now becoming at parity with doctors, lawyers, and hedge fund analysts, which will hopefully equate things."

HOW CAN US SCIENTIFIC AND TECHNOLOGICAL LITERACY BE IMPROVED?

As science and technology become ever more entwined with major public policy issues, the public needs to engage in policy discussions that have a scientific and technological component. But current levels of literacy in science and technology are not high enough to have these discussions, said Katehi. "I don't believe we have the ability to engage in science discussions and talk about nuclear energy, GMOs, . . . even Google and how it is constructed and utilized. We don't have the ability as a public to be engaged meaningfully."

One way mentioned by several speakers to build scientific and technological literacy among college students is to bring together students from different fields and have them work together. For a design problem, for example, students from the humanities, the arts, business, and engineering can all make distinctive contributions. "Design is not just an engineering skill. It's a skill for everyone," said Katehi.

> "Design is not just an engineering skill. It's a skill for everyone."
> *Linda Katehi, University of California, Davis*

Building on this idea, Teeri described design in the context of the fashion industry. Aalto University had a fashion show at the beginning of a recent term to emphasize the elements of design that go into fashion. Cocreation platforms and teamwork can build the kind of literacy with technical subjects that students need. "If you are a fashion designer and you see that there is a materials scientist who is working with something interesting for you, then obviously it becomes interesting and less difficult, because you have the motivation to learn about it."

Engineers also could be encouraged to talk about engineering with students from other fields, suggested Miller, just as music students put on recitals for nonmusicians. Such performances could help engineering students become better communicators while helping nonengineers understand what engineers do.

For example, Aalto University has taken the idea of a Masters of Art exhibition from the design school and turned it into a Masters of Aalto exhibition, where master's-level engineers and designers showcase the products of their graduate work. "They get ideas from each other, but not least, they get to meet potential future employers. So it's not only an exhibition of their work, but it also turns into a kind of job fair at the same time."

WHAT SHOULD BE THE ROLE OF TEACHERS?

In an online world, said Agarwal, the most valuable role for teachers is to be a guide for the development of communities and the discussion of ideas. Because of their online experiences, today's students are

NAE member Jennie S. Hwang participating in a group discussion.

very comfortable typing questions into electronic devices and reading responses. For the edX forums, just a handful of teachers were able to support the thousands of students taking the course because students usually answered each other's questions—and were "learning as they did." As a result, students became teachers for each other.

Miller agreed that student engagement with faculty members will remain essential, even as online learning "changes the game." Communication has many different forms and many different purposes. If all communication occurs through texting, essential skills will be lost. "You need to be there in person. You need to make eye contact. You need to develop confidence in who you are as a person, and trust."

"We are going to have a whole portfolio of ways of learning, and people will be commuting between the different ways."

Tuula Teeri, Aalto University

Everyone has a different set of strengths and weaknesses, said Teeri. As a result, different kinds of teaching and learning will appeal to and best serve different people. Some people will prefer online learning, while others will opt for more practice-based learning. "We are going to have a whole portfolio of ways of learning, and people will be commuting between the different ways."

Katehi agreed that online interactions cannot replace human interactions. "I don't believe that anyone can learn without a teacher playing a key role," she said. "Social scientists will tell you that there is a human need to be connected and to come into contact with and learn from others. It's as if we would believe that we can produce professional football players just by learning from each other without a coach. We would not have any football teams. That's the role of the teacher, the role of the coach."

Khan also agreed that online learning will never replace human interaction, and particularly the inspiration and wisdom that mentors can deliver. But the possibilities for online learning are just starting to be explored and will continue to diversify and develop. Already, online learning raises the question of why students continue to learn from 300-person lectures. A revolutionary outcome, he said, would be for online learning to replace lectures on college campuses within five years. Instead, learning could be interactive and project based, with students teaching each other and professors providing mentoring and guidance.

WHAT SHOULD BE THE ROLE OF TECHNOLOGY IN EDUCATION?

While technology has great potential to change education, some panelists argued for more time making things rather than interacting with screens. In recent decades, US children have spent less time outside and in rural areas where they can become engaged with the physical world, Stephens said. "They are more digitally capable, but less real world–savvy." For example, when Steve Jobs was growing up, Stephens observed, he was taking apart and reassembling cars, not computers.

Katehi stressed the importance of engaging children very early in real-life problems. "A lot of kids want to change their lives. They want to change the world." Traditional engineering education provides students with large amounts of content; then, toward the end of their educational experiences, they get to start designing things. The order should be reversed, said Katehi. "We need to start with practice, get them to experience the impact of their solutions—even if they may be incomplete or suboptimal—and get them excited in what they are doing. That's what is going to get more kids to come to engineering and is going to make the engineering workforce more diverse." Online experiences are a useful supplement to hands-on experiences, but people need to know how to ask questions, search together for solutions, and take satisfaction in solving problems.

> "A lot of kids want to change their lives. They want to change the world."
>
> *Linda Katehi, University of California, Davis*

Speaking as an employer of engineers, Khan agreed with the need for more problem solving in education. Rigor in engineering education is useful, but if not done properly it can squeeze out creativity, he said. He often talks with students from top engineering schools, but when he asks them what they have built, he finds that they have simply done problem sets and projects that were already well defined. "That's one of our main filters for people: if they ever created anything."

Khan described one of his strongest employees as the "world's best JavaScript programmer," the author of a JavaScript library known as jQuery that is used by millions of people. He was hired on the spot during his interview. But when Khan asked him over lunch two months later what his grade point average was, he said, "Oh, I think it was 1.9."

"I was blown away," Khan recalled. "We would never have interviewed someone with a 1.9 GPA. I [said], 'You're the smartest guy I know.' He said, 'Well, I was working on jQuery the entire time.'"

Another employee, one of the company's best developers of online exercises, was a college dropout. "He just couldn't get engaged. He liked to create things. He didn't like to sit and take tests." Stressing the creativity inherent in engineering, and not just the rigor, is one of the best ways to get more people interested in engineering and more students graduating from engineering programs, Khan insisted.

HOW CAN MORE CHILDREN BECOME INTERESTED IN ENGINEERING?

Differences of opinion arose in response to a question from the audience about how to encourage young children to become interested in engineering. Make sure they play with Legos, advised Khan. "I don't know if you all have seen some of these new Lego kits with the Mindstorms. You could do a PhD thesis with these things. They have heat sensors and touch sensors and light sensors and memory. I have seen 9-year-olds do fantastic, creative, really deep things."

Khan also said that he would teach his young children to program a computer by the time they are ten. Some Khan Academy lessons are aimed at children below the age of ten, and even very young children are able to master sophisticated concepts when they see learning as fun. "It's amazing what kids can do if you approach the same content in the form of play as opposed to a didactic, mathematics type of thing," he said.

Stephens countered that the best way to create future engineers is to "get them outside, get them to go play, go interact. They will get the computer skills when the time is right." He said that it is more important for young children to get hands-on experience in the real world. Academic experiences can then help children understand their experiences. "In my mind, it's about experiences, it's about playing, it's about involving. Those really build the social skills."

> "It's amazing what kids can do if you approach the same content in the form of play as opposed to a didactic, mathematics type of thing."
>
> *Salman Khan, Khan Academy*

For girls, Katehi mentioned the messages they get from their mothers. When they are very young, children tend to spend time with their mothers at home or with other caregivers who tend to be women.

"It's critical to make sure that our girls, who will become mothers in the future, have enough understanding and appreciation of science and engineering so they can teach it to their children." The adults in the lives of children need to encourage curiosity, finding the answers to questions, and other precursors to an engineering perspective.

Children are interested in big problems and grand challenges. They care about sustainability, human health, and personal security. Miller said that children need to learn that they have the potential to make a difference in the world, despite its size and complexity. "It's about envisioning things that you can do, even on a small scale, to build a can-do attitude." Say a child was to see a classmate struggling to get on or off the school bus. That is a problem to solve. "If they succeed at that, it will exceed their expectations. And if they do that three times in a row, they begin to have a can-do attitude," Miller said.

Teeri agreed that engineering education starts too late. When children are young, they have no idea what scientists or engineers do. "They see more lawyers than engineers on TV." Teeri described an event at Aalto University where preschool children came to the university, based on research that children exposed to a university at an early age are more likely to pursue academics than children who have not been to a

Anant Agarwal and Charles M. Vest in conversation.

university. "We were asking them if they knew what a university is, and one of them said, 'A university is a place where you can become what you want to be.' I can't think of a better definition than that."

Children spend many hours a day with electronic media, Stephens observed. But very few characters they witness in these media are related to science, technology, engineering, or mathematics—and those who are tend to be portrayed as villains or fools. Industry has begun to work with the entertainment industry to change these stereotypes, and the entertainment industry is receptive, partly because it, too, needs the help of scientists and engineers to be successful. For example, the SET Awards—for science, engineering, and technology—are presented to movies, TV series, radio and TV news programs, and print and online journalism for accurate and impactful entertainment portraying and promoting the fields of science, engineering, technology, and mathematics. "They have grabbed on to this and said, 'We have to help solve this problem as well.'"

Khan made an intriguing suggestion about how to pull the humanities and engineering closer together. Today, students in high school English classes typically study great literature from historical periods such as the Victorian era. But almost completely absent from K–12 English classes is science fiction. Many science fiction novels juxtapose ideas from science and engineering with ideas from the social sciences, implicitly drawing those fields closer together. And in contrast to the literature usually studied in English classes, science fiction is forward looking. Some of the best science fiction "is about extrapolating science to what it can be."

THE COMING CHANGE

NAE president Charles Vest closed the forum by thanking the speakers and calling attention to how quickly the world is changing. The new world of engineering education is "coming down the pike so fast that it's going to hit everybody like a ton of bricks," he said. "We would be remiss if we underestimated the role that information technology is going to play. We are just at the beginning. We don't know where it's going to go, but it's going to be big,...and it's very exciting, because we are going to create opportunity for vast new numbers of young women and men."

> "We are going to create opportunity for vast new numbers of young women and men."
>
> *Charles Vest, NAE*

Appendix A

Forum Agenda

Annual Meeting Forum

**Educating Engineers:
Preparing 21st Century Leaders in the Context of New Modes of Learning**

Monday, October 1, 2012
9:30 a.m.–12:30 p.m., Eastern Daylight Time
National Academy of Sciences Building
Washington, D.C.

Engineering education is changing in the United States and around the world. Forces include integration with new science, particularly nanoscale science and biology; increased emphasis on innovation and entrepreneurship to drive economic growth; accelerating speed, complexity, and globalization of engineering functions; preparation of graduates to address the Grand Challenges; empowering a more diverse population of young American engineers; renewing a commitment to design and production; and developing new approaches to large-scale engineering systems that must engage society. As in all of higher education, there also are vast challenges such as declining financial support and equally vast opportunities such as the approaching tsunami of interactive online learning and massively open online courses.

What lies ahead? What are the ramifications for engineering schools? In this year's Forum, Educating Engineers: Preparing 21st Century Leaders in the Context of New Modes of Learning, an expert panel will explore many facets of these challenges and opportunities and offer guidance on the roles and responsibilities of 21st century engineering educators.

Welcome

Charles M. Vest, President, National Academy of Engineering

Moderator: Ali Velshi, Anchor and Chief Business Correspondent, CNN

Forum Discussion

Forum Participants:

Anant Agarwal, President, edX and Professor, Massachusetts Institute of Technology

Linda P. Katehi, Chancellor, University of California, Davis

Salman Khan, Founder of the Khan Academy

Richard K. Miller, President of Olin College

Richard (Rick) Stephens, Senior Vice President for Human Resources and Administration, The Boeing Company

Tuula Teeri, President, Aalto University, Finland

Appendix B

Biographical Information

ALI VELSHI is CNN's chief business correspondent and host of *Your Money*, CNN's weekend business roundtable program, as well as a regular contributor and anchor for "Issue #1," the network's in-depth coverage initiative on the single issue that matters most to CNN's audience. He also hosts *The Ali Velshi Show*, a weekly call-in radio program on both CNN Radio and CNN.com Live and fields viewer calls three times a week for the "Help Line" segment for HLN. Online users can listen to Mr. Velshi's podcast, "The Ali V Podcast," available at www.CNN.com/podcasting and on iTunes.

Based in New York, Mr. Velshi has covered the US government's bailout plan; the financial collapses of Fannie Mae, Freddie Mac, AIG, and Lehman Brothers; and Hurricanes Gustav and Ike, reporting on the impact of the storms on oil refineries. He covered the Enron story at every step when it hit the national spotlight in 2001, including the guilty verdicts of Enron Corporation's founder Kenneth Lay and former chief executive Jeffrey Skilling on conspiracy and fraud charges. He reported live from Ford headquarters in Dearborn, Michigan, as the company announced the layoff of 30,000 workers. He was reporting live from an oil rig in the Gulf of Mexico amid evacuation calls for Hurricane Katrina.

A veteran of financial news, Mr. Velshi recently hosted *The Turnaround*, CNN's small business improvement show. *The Turnaround* traveled across America, introduced troubled small business owners to high-profile mentors, and then helped them develop a plan for success.

Before *The Turnaround*, Mr. Velshi was an anchor with CNNfn, where he hosted various interactive shows, including *Your Money, Busi-*

ness Unusual, Insights, Street Sweep, and *The Money Gang*. Before joining CNNfn in 2001, he hosted *The Business News*, Canada's first and only prime-time business news hour, airing nightly on Report on Business Television.

Earlier in his career, Mr. Velshi worked as a business anchor for Cable Pulse 24 and its sister station, CITY TV, and as a reporter for CFTO-TV, Canada's most-watched local television station.

In 1996, Mr. Velshi was awarded a fellowship to the US Congress by the American Political Sciences Association and worked with now-retired US Representative Lee Hamilton (D-Ind.).

Born in Kenya and raised in Toronto, Velshi graduated from Queens University in Kingston, Ontario, in 1994 with a degree in religion. Mr. Velshi's first book, *Gimme My Money Back: Your Guide to Beating the Financial Crisis*, was released in January 2009. He is a member of the Grand Challenges Advisory Committee for the National Academy of Engineering, the Economic Club of New York, and the New York Financial Writers Association.

ANANT AGARWAL is the President of edX, a worldwide online learning initiative of MIT and Harvard University, and a professor in MIT's Electrical Engineering and Computer Science Department. He has also served as the Director of MIT's Computer Science and Artificial Intelligence Laboratory (CSAIL). He leads the Carbon group,

which focuses on research involving operating systems and architectures for manycores and clouds. He is also a founder and CTO of Tilera Corporation, which created the Tile multicore processor. Dr. Agarwal holds a PhD from Stanford and a bachelor's from the Indian Institute of Technology, Madras. He led the development of Raw, an early tiled multicore processor; Sparcle, an early multithreaded microprocessor; and Alewife, a scalable multiprocessor. He also led the VirtualWires project at MIT and was the founder of Virtual Machine Works, which took the VirtualWires technology to market. He is an author of the textbook *Foundations of Analog and Digital Electronic Circuits*.

LINDA P. KATEHI became the sixth chancellor of the University of California, Davis, in August 2009. As chief executive officer, she oversees all aspects of the university's teaching, research, and public service mission.

Chancellor Katehi also holds UC Davis faculty appointments in electrical and computer engineering and in women and gender studies. A member of the National Academy of Engineering, she chaired until 2010 the President's Committee for the National Medal of Science and the Secretary of Commerce's committee for the National Medal of Technology and Innovation. She is a fellow of the American Association for the Advancement of Science and the American Academy of Arts and Sciences and is a member of many other national boards and committees.

Previously, Chancellor Katehi served as provost and vice chancellor for academic affairs at the University of Illinois at Urbana-Champaign; the John A. Edwardson Dean of Engineering and professor of electrical and computer engineering at Purdue University; and associate dean for academic affairs and graduate education in the College of Engineering and professor of electrical engineering and computer science at the University of Michigan.

Since her early years as a faculty member, Chancellor Katehi has focused on expanding research opportunities for undergraduates and improving the education and professional experience of graduate students, with an emphasis on underrepresented groups. She has mentored more than 70 postdoctoral fellows, doctoral and master's students in electrical and computer engineering. Twenty-two of the 44 doctoral students who graduated under her supervision have become faculty members in research universities in the United States and abroad.

Her work in electronic circuit design has led to numerous national and international awards as both a technical leader and educator, and 19 US patents. She is the author or coauthor of 10 book chapters and about 650 refereed publications in journals and symposia proceedings.

She earned her bachelor's degree in electrical engineering from the National Technical University of Athens, Greece, in 1977, and her master's and doctoral degrees in electrical engineering from UCLA in 1981 and 1984, respectively.

SALMAN KHAN is the founder of the Khan Academy (khanacademy.org), a nonprofit with the mission of providing free, high-quality education for "anyone, anywhere" in the world. He graduated from MIT in 1998 with three degrees: two bachelor of science degrees in mathematics and electrical engineering/computer science, and a master of science degree in electrical engineering. He worked in technology in Silicon Valley until the first bubble burst, after which he attended Harvard

Business School. After earning a master's degree in business administration in 2003, he became an analyst at a Boston-based hedge fund. In 2004 as a side project, Mr. Khan began tutoring his young cousin in math, communicating by phone and using an interactive notepad. When others expressed interest, he began posting videos of his hand-scribbled tutorials on YouTube. Demand took off, and in 2009 he quit his day job to commit himself fully to the not-for-profit Khan Academy.

In October 2010, Mr. Khan was listed in *Fortune's* annual "40 Under 40," which recognizes business's hottest rising stars, as well as *Fast Company*'s list of the "100 Most Creative People in Business." He was recently profiled by "60 Minutes" and recognized by *Time* magazine as one of the 100 most influential people in the world.

He was born in Metairie, Louisiana, to immigrant parents from India and Bangladesh. He currently lives in Mountain View, California, with his wife and two children. *The One World Schoolhouse: Education Reimagined* is his first book, published in October 2012.

RICHARD K. MILLER was appointed president (and first employee) of the Franklin W. Olin College of Engineering in 1999, where he also holds an appointment as professor of mechanical engineering. He served as Dean of the College of Engineering at the University of Iowa from 1992 to 1999, and spent the previous 17 years on the engineering faculties at the University of Southern California and the University of California, Santa Barbara. Dr. Miller's research interests are in innovative engineering education and applied mechanics, on which he has authored about 100 publications and given numerous keynote presentations. He is a member of the National Academy of Engineering, has won five teaching awards, and received the Legacy Award from the College of Engineering at the University of Iowa. He has also served as chair of the US National Science Foundation's Engineering Advisory Committee and served on advisory committees for the World Bank, National Academy of Engineering, Harvard University, and other institutions. Dr. Miller earned a BS degree in aerospace engineering from the University of California, Davis, from which he received the 2002 Distinguished Engineering Alumnus Award. He earned an MS in mechanical engineering from the Massachusetts Institute of Technology and a PhD in applied mechanics from the California Institute of Technology.

RICHARD (RICK) STEPHENS, a 32-year veteran of the Boeing Company, is senior vice president, Human Resources and Administration, and is a member of the Boeing Executive Council.

Mr. Stephens serves on a number of nonprofit and business-focused boards and has been recognized for his longstanding leadership in local and national organizations. Passionate about improving education both in and out of the classroom, he works directly with community and education leaders to prepare future workers to meet the challenges necessary to succeed in an ever-changing and competitive business environment.

Mr. Stephens currently serves on the National Governors Association Center for Best Practices Science, Technology, Engineering and Math (STEM) Advisory Committee, designed to assist governors in developing comprehensive STEM agendas. He also is a Fellow of the American Institute of Aeronautics and Astronautics where he is chair of the Career and Workforce Development Steering Committee, a member of the Business-Higher Education Forum, and founding member of the Business and Industry STEM Education Coalition (BISEC). Earlier, Mr. Stephens served on the Secretary of Education's Commission on the Future of Higher Education, the President's Board of Advisors on Tribal Colleges and Universities, the National Academies Committee on Science, Engineering, and Public Policy, and the Association of Public and Land-Grant Universities Science and Mathematics Teacher Imperative Commission. These diverse and related education experiences, along with his leadership in a major technology-based company, give him unique insights into how education can prepare students to be successful in the future job market.

TUULA TEERI was appointed the first president of Aalto University in April 2009. She is a staunch supporter of research excellence as a solid foundation for pioneering education and societal impact. Under her leadership, Aalto has implemented the tenure track career system, established interdisciplinary programs in research and education, and is reforming its educational curricula according to the Bologna principles. Aalto University is also emerging as a major hub for growth entrepreneurship through joint activities of the student-led Aalto Entrepreneurship Society, Aalto Venture Garage, and Aalto Center for Entrepreneurship.

Dr. Teeri earned an MSc degree in molecular genetics in 1981 and a PhD degree in 1987 from the University of Helsinki. She held various research and leadership positions at VTT Technical Research Center of Finland during 1980–1996. She then joined the Royal Institute of Technology (KTH) in Stockholm as a Visiting Professor (1996–1998) before her appointment as Professor in Wood Biotechnology in 1998. Her research career focused on enzyme discovery and protein engineering of cellulose degradation and plant cell wall biosynthesis with long-term aims of developing processes for biofuel applications and for new biomimetic materials.

At KTH, Dr. Teeri served as a member of the Faculty Board (2003–2007), Deputy Dean of the School of Biotechnology (2005–2007), and Deputy President of the University (2008–2009). She has been a board member of the Institute of Surface Chemistry, Sweden (2001–2004), Institute for Future Studies, Stockholm (2003–2006), Council for Natural Sciences and Technology of the Swedish Research Council (2004–2006), and Swedish Foundation for Strategic Research (2004–2007). She is a current board member of the Technology Academy of Finland, Finnish Business and Policy Forum, and the Research Institute of Finnish Economy, Universities Finland UNIFI, as well as the European Science Business Board.

Tuula Teeri is a member of the Royal Swedish Academy of Sciences, the Royal Swedish Academy of Engineering Sciences, Technology Academy Finland, and the Swedish Academy of Technology in Finland. She is also cofounder of SweTree Technologies.